FOREWORD

The Iceland Horse is not sufficiently recognized in North America, and I hope that this book will help to create a better understanding of this remarkable breed. It first came to Iceland with the arrival of the pagan Norsemen who settled Iceland in the period 870—930 A.D. Since then, there has been some admixture of other breeds, but its admirable qualities have not been diluted, which has happened in so many other cases.

These qualities include strength, endurance, gentleness, character, keen vision, and a good sense of direction. The Norwegian Northland horse, which is virtually unknown outside Norway, is in some ways similar to the Iceland Horse.

Horse or pony? It depends on your definition. The European association for the breed is called the "Federation of European Friends of the Iceland Horse." The Belgians, Dutch and French say Iceland Pony. The Norwegians, Swedes, Danes, Germans and Swiss refer to the Iceland Horse, as do the Icelanders. I have assumed the Icelandic publisher's prerogative and call it the Iceland Horse. Technically, a height of 14.2 hands or under is considered "pony," and here the Iceland Horse qualifies. But take another excellent breed, the Connemara "pony" from the West of Ireland (which my family raises and rides). When this animal leaves its native habitat for literally greener pastures, it frequently grows to fifteen hands and more. Horse or pony? There is a presumption that there is a marked difference, and not only in size, between the two categories.

The text of this book is by Sigurdur A. Magnusson, an Icelandic man of letters and former president of the Icelandic Writers' Union. He possesses the added virtue of having ridden since early childhood. He writes of the important role of the horse in Icelandic mythology, literature and history, as well as its breeding, care, and accomplishments in recreational riding and in equestrian competition.

THE PUBLISHER

STALLION
OF THE NORTH

Text: Sigurdur A. Magnússon
Photos: Gudmundur Ingólfsson and others

Longship Press

Photos:	Page
U. Becker	92, 93, 94, 95
Kr. Benediktsson	45, 76, 77, 78, 79, 80, 81, 82, 83
G.B. Björnsson	41, 91
M. Chillmaid	8, 23
G. Hannesson	3, 5, 7, 9, 11, 24
G. Ingólfsson	Front cover, 16, 17, 20, 27, 28, 29, 30, 31, 33, 34, 35, 36, 37, 38, 44, 47, 48, 49, 50, 51, 52, 53, 54, 56, 59, 60, 61, 62, 64, 65, 66, 68, 69, 70, 73, 74, 75, 84, 85, 86, 87, back cover
E. Isenbügel	15, 89, 90
M.W. Lund	18
B. Samper	13, 39, 40, 43, 46
F. Thorkelsson	55, 57, 58, 62, 63, 88

Published in 1978 by Iceland Review ©, Reykjavík, Iceland
and Longship Press, Crooked Line, Nantucket, Mass. 02554.
ISBN — 0-917712-06-4
Library of Congress — 78-67094
Printed in Italy

Divine Origin

For some not so hidden reason, the horse has been a stirrer of human imagination since he was first tamed and taken into man's service some 6,000 years ago, and perhaps even longer. Among all Indo-Germanic nations, the horse was an integral part of daily life, religion and art. We have references to horses in various mythologies, most famous of which are the winged Pegasus of Greek mythology (sprung from the Gorgon Medusa's blood) and Sleipnir of Norse mythology (progeny of the evil Loki when he once turned himself into a mare), the eight-footed steed of Ódin, the principal god. Why the origins of these two most famous of horses should be connected with evil parentage is a dark mystery.

There are other renowned horses in Norse mythology. One of Ódin's sons, the keen-sensed Heimdall, the watchman of the gods, who could hear the growing of grass on the ground and of wool on the backs of sheep, had a horse named Goldtop. The goddess Gná, daughter of Frigg (Ódin's wife), had a horse named Hófvarpnir ("one who tossed his hoofs"), who raced through the air and over the waters. The white and pure god Balder's horse is not mentioned by name, but it accompanied

its master on his last journey, to the funeral pyre. Several horses of the gods are mentioned by name without further identification: Gladur, Gyllir, Glenur, Seidbrúnir, Silfrintoppur, Sinir, Gísl, Falhófnir, Léttfeti. Ódin, the chief god, was notorious for his numerous disguises and pseudonyms, many of which were the names of horses, for instance Jálkur, Fengur, Vakur, Brúnn, Hrani, Hrosshársgra-

ni, etc. The world-ash tree, symbol of the life-force, where the gods held their solemn assemblies, was called Yggdrasil, "Ódin's horse."

The many equine appellations of Ódin have by some authorities been taken to indicate that he was originally a primitive deity, worshipped in the form of a horse. Still more primitive was the worship of trees and groves. Later those two merged, and the horse was worshipped in sacred groves. In consequence, the place where the gods assembled was named the Ash of Yggdrasil, the ash of the awe-inspiring god in horse-guise.

There are also clear indications that the horse was worshipped as a deity or symbol of fertility. He was in many cases dedicated to Frey, the god of fertility, which was a later development, but the whole procedure and content of the *blót* or sacrificial feast shows the religious significance of the horse. When the priest (*godi*) had opened the feast and declared the truce, a white horse with a silver-gray mane was led forward. It had never been used for any non-sacred work. The priest slaughtered it either by sticking a double-edged iron into the nape of the neck and cutting the spinal cord or by breaking the skull with an axe which was not used for other purposes. The horse's blood was ga-

3

thered in a bowl, an ancient stone vessel in which the sacrificial blood was kept, but while this was being done the priest consecrated the blood and seized a bunch of twigs, which was probably originally the tail of the sacrificial beast, and carried the blood through the place of worship, scattering it on the altars, on the gods on the altars and on the walls inside and outside. The blood was also sprinkled on the congregation, and those most ardent in their religious fervour drank of it. The next stage of the feast was to drink to Njörd and Frey, the gods of the chase, of good fortune at sea and of fertility, and to Ódin. Men drank to good fortune, peace, prosperity and concord. The last stage was to partake of the meat and liver of the sacrificial horse in order to become one with the deity or acquire some of its qualities, and to reinforce the deity itself as it was resurrected after the sacrificial feast.

When Christianity was adopted in Iceland in the year 1000, one of the principal stipulations was a prohibition against eating horse-meat. In a sense, the fight for Christianity in the North was a fight against the horse, against its worship and the eating of horse-meat. In some parts of Iceland, the fear or abhorrence of this heathen practice has struck such deep roots

that to this very day people refuse to touch horse-meat, while in other parts it is relished. The writer knows several people who have literally gone out and vomited when informed that a week earlier they had unwittingly eaten horse-meat.

Giant Horses
The giants, arch-enemies of the gods, had horses just like the gods and the human race. There was the very fine horse named Goldmane, one that the thunder-god Thór won from them in a famous duel, with the aid of his three-year old son Magni. Then there is the amusing story of how Sleipnir came into being: When Midgard, abode of the gods, had been created and the gods were meditating the building of a massive stronghold as a bulwark against the giants, a giant smith came forward and offered to build the stronghold in a year's time if he might have the goddess Freyja, the sun and the moon by way of payment; but if on the first day of summer any part of the work remained undone, he was to receive no wages. The building proceeded more rapidly than they had thought possible, for the giant's powerful horse, Svadilfari, during the night pulled into place stones as huge as mountains. When

only three days remained before the arrival of summer, the giant was already busy with the castle gate, and the gods were growing uneasy. They summoned Loki whose bad counsel was the cause of their trouble, threatened him with death, and thus frightened him into promising to find a way out of their difficulties. Transforming himself into a mare, he ran whinnying out from the forest at evening just as Svadilfari was at his task of hauling stones. Svadilfari broke loose and followed the mare into the woods, pursued in turn by the builder; that whole night not a stone was hauled, and thus the work was interrupted. The mason was enraged; but Thór crushed his head with his hammer Mjöllnir. The mare — or Loki — later foaled Sleipnir.

The divinities of Day and Night both of giant origin, were given two horses and two wains by Ódin, who placed them aloft in the heavens where they were to ride around the earth in alternating courses of twelve hours each. Night drove the horse known as Hrímfaxi ("having a mane of rime"), and each morning the fields were bedewed with the froth that dripped from his bit. Day drove Skinfaxi ("with the shining mane"); earth and sky sparkled with the light from his mane.

There is another myth about the celestial bodies being created by Ódin and his two brothers, the first gods. To this end they employed the sparks that flew from Muspellsheim. The sun and the moon were placed each on its wain, and each wain was drawn by two horses; the horses of the sun were named Árvakur ("the early-waking one") and Alsvin ("the fleet one"). Before the sun stood the shield Svalin ("the cooling one"). As drivers of the wains were appointed the two beautiful children of Mundilfari, called Sun and Moon. Mundilfari was so proud of the two that he had named his daughter for the sun and married her to a man named Glenur ("the bright one"), and his son for the moon. As a punishment, the gods gave the children the task of guiding the wains of the sun and the moon.

If we consider that the husband of the sun carries the name of a horse (Glenur) and that Loki, the son of the fire and the evergreen tree, turns himself into a mare three days before the arrival of summer and staves off the danger posed to the sun by the first day of winter, and also that the dew of fields and the light of earth and sky are traced to Hrímfaxi and Skinfaxi, it seems more than likely that these myths are connected with ancient customs rooted in the worship of the horse in order to enhance the fecundity of the earth. The story of how Sleipnir came into being would then be an explanation of rites used to reinforce the horse of the sun in a new resurrection. This was probably done with some kind of festivities, and the First Day of Summer, Iceland's only native holiday still observed, may well be the last vestige of that festival.

Legendary Heroes and Horses

The Eddic texts preserved in Iceland are virtually the only pre-Christian Germanic literature now extant, preserving not only the mythology of our ancestors, but also the oldest heroic legends of the North, those of Sigurd (Siegfried) and Helgi the Hunding-Slayer. The heroic poems relating the Sigurd legend contain the greatest tragic story of Western Europe and one of the most sensitive descriptions of the relations between man and horse. The story of Sigurd and his horse Grani has been recited and sung throughout Western Eruope for more than a millennium: Sigurd, who was of royal blood, lost his father in childhood and was brought up in servility. He excelled in all things, physical as well as mental, was a true hero, courageous and blameless, but his childhood experience had deeply affected him, divided his soul, deprived him of resolution at decisive moments when bold action was needed. That proved to be his undoing much as it was Hamlet's. Even if Sigurd excelled in every respect, his greatest endowment was his horsemanship and his priceless horse, Grani, the like of which was never found on earth.

The first task set before Sigurd when he had been made aware of his origin was to choose himself a horse from a famous king's stud. He chose the light-gray Grani, the horse that made him what he became and was a party to all his great deeds, including that of killing the terrible dragon Fáfnir and taking all its gold. We are told that he filled two huge chests with gold and bound them to a pack saddle on Grani's back one on each side; he meant to drive the horse before him, but Grani would not move a foot before Sigurd himself mounted. After this mighty deed, Sigurd fell in love with Brynhild and even had a child by her, but Queen Grímhild, mother of Gudrún, employed magic to make him forget Brynhild so that he would marry her daughter, which he did. His brother-in-law, Gunnar, desired Brynhild and they went to woo her. She had made known her intention to marry only him who would ride through a ring of flames surrounding her hall. Gunnar's horse, Goti, recoiled, so Sigurd made him a loan of Grani, who refused to stir a pace. Sigurd and Gunnar therefore had to change guises and names, and Sigurd rode Grani through the flames disguised as Gunnar. Sigurd then wedded Brynhild but laid his naked sword between them during the eight nights he spent in her hall. He then returned and changed guises with his brother-in-law. Some time later, a quarrel between Gudrún and Brynhild over the relative merits of their husbands revealed the truth, and Brynhild did not relent until she had caused Sigurd to be killed, after which she herself followed him in death, her last wish to be burned with him on the funeral pyre. That wish was granted, and we may assume, even though it is not mentioned, that Grani followed them along with their entourage, five bond-women and eight henchmen. In one sense, it was

A stallion in summer, alert and
noble, imperious and savagely
jealous of his dominion. A paragon of strength and beauty.

From the Thing ran Grani
with thundering feet,
but thence did Sigurd
himself come never;
covered with sweat
was the saddle-bearer,
wont the warrior's
weight to bear.

Weeping I sought
with Grani to speak,
with tear-wet cheeks
for the tale I asked;
the head of Grani
was bowed to the grass,
the steed knew well
his master was slain.

A Roman Account

The preoccupation with horses in the myths and heroic legends of the North leads us back to the Roman historian Tacitus, who in 98 A.D. gave an account of the nature and manners of the Germanic peoples in his *Germania*. According to him, our forefathers north of the Rhine did not live in towns, but scattered, each family together, wherever there was water or a pretty grove. The food was plain, consisting of fruit, fresh venison and curdled milk. They were also eager gamblers, throwing dice when quite sober, as if that were honest work. They had good horses and their

horsemanship was renowned, although it is not described; but he states, that it was the pastime of children, the aim of young men, and also much indulged in by old men.

In Chapter X of his book, Tacitus has this to say about the relations of our ancestors with their horses: "But what is unique about this nation is that it tries to obtain omens and predictions from horses. The animals are reared at public expense in the afore-mentioned sacred woods. The horses are snow-white and have never been used for any non-sacred work, and when they have been harnessed to the divine chariot they are accompanied by the priest, the king, or the head of state, who studies their neighs and snorts. No oracle is more sacred than these horses, not only among the populace but also among the chieftains, for although the Germans regard the priests as the servants of the Lord they regard the horses as his confidants."

In the light of later developments, it is interesting to note that another Roman author, General Gallus, writing in 66 A. D., has this to say about the Arabs: "The fortunate Arabs own a great number of cattle, but they have neither horses nor mules nor hens nor geese." The noble Arabic breed, so important in improving

Goti's fear of the fire and Grani's courage that had sealed their destiny.

In her old age, Gudrún recalls those tragic events, and her mind dwells especially on Sigurd and his horse. The intensity of the friendship between man and horse and the regret of the animal are beautifully rendered in these ancient stanzas:

The studhorses are usually full of play and vitality in spring and early summer, their bodies slim and vibrating with life, their necks raised, their ears pricked up, their eyes clear and watchful.

other breeds (with the exception of the Iceland Horse), originally came from Persia and Mongolia and did not come to the fore until the early seventh century, during the latter part of Mohammed's life (570–632).

Saga References

In the *Sagas of Icelanders,* that unique body in world literature, there are numerous references to horses. One of the very best Sagas, the concise *Hrafnkels's Saga* from about 1270, has a horse as the central motif, underscoring that the horse played an important role in the religious cult of Iceland before the year 1000. The horse, Freyfaxi, had been dedicated to the god of fertility, Frey, by his owner, who forbade anyone to mount him, at the risk of forfeiting his or her life. A servant disregarded the order, and was as a consequence killed by his master. That event brought about the dramatic action of the Saga, in the course of which the horse was pushed off a high cliff into a river. That episode brings to mind the kelpie, the malevolent horse living in rivers, which is so common in Icelandic folklore.

It is evident from other sources that riding a horse consecrated to the gods was considered a heinous crime, and according to the oldest Icelandic lawbook, *Grágás,* stealing horses was a capital crime, resulting in outlawry or banishment from the country. There is an extensive section on the subject in that book, but it was omitted from the lawbook of 1280, *Jónsbók,* which was largely in force until the nineteenth century. The omission probably reflected the fact that the sacredness of the horse had greatly diminished by the end of the thirteenth century. However, stealing horses continued to be a very serious crime, and in 1786 a royal letter was issued stipulating that horse-thieves were to be flogged and condemned to lifelong slavery in Copenhagen, the men to carry chains.

Because of the magic powers attached to horses in ancient times the so-called scorn-poles were put up to cause one's enemies' downfall or ridicule. The most famous example in the Sagas is that of Egil Skallagrímsson, the great poet-Viking, erecting a scorn-pole against his arch-enemies, Queen Gunnhild and King Eirík Bloodaxe, finally driving them out of their kingdom. The incident is described in *Egils's Saga:* "He (Egil) took a hazel pole in his hand and went to a certain jutting rock facing the mainland. Then he took a horse's head and set it up on the pole. Afterwards he recited a formula, saying these words: 'I here set up a scorn against King Eirík and Queen Gunnhild.' He turned the horse's head landwards. 'I turn this scorn upon the landspirits which dwell in this land, so that they all fare wildering ways, and none light on or lie in his dwelling till

9

they drive King Eirík and Gunnhild out of the land.' Next he jammed the pole down into a crack in the rock and let it stand there. The head he turned landwards, but he graved runes on the pole and they state all that formula." The royal couple's misfortunes and flight from Norway almost came as a natural consequence of this undertaking, according to the Saga, written more than two centuries after the adoption of Christianity.

There are many references to horses being sent from Iceland to Norway and elsewhere as gifts to kings and other notables. As a matter of fact, horses were considered the very finest of gifts, along with hawks and gyrfalcons. This tradition has been preserved in Iceland until modern times. In 1874, when the Icelanders celebrated the millennary of the settlement and the first Danish king visited Iceland, being also the king of Iceland, a farmer from the south of Iceland brought him his best riding horse as a present, as a result of which he was invited to visit the royal family in Copenhagen. The farmer made the trip and wrote a book about his experience, which is now a classic in Icelandic letters. This same farmer is the hero of Halldór Laxness's celebrated novel, *Paradise Reclaimed.* When Queen Margaret of Denmark came to the throne in 1972, she and her husband were sent two selected horses by the Icelanders, in accordance with ancient traditions.

Races

Horse-races are mentioned in our ancient records and so is betting, but there are no references to organized races such as those that seem to have been practised in Britain and elsewhere at the time. In the *Book of Settlements* (*Landnámabók*) there is an account of a fine mare that was transported from Norway and tried in a memorable race in the uninhabited part of Iceland. It runs like this: "At that time a ship came to Kolbeinsarós laden with livestock, but they lost a mare into Brimnes Forest which Thórir Pigeon-nose bought, if it should ever be found, and later he found it. It was the swiftest of horses and called Fluga (Fly). A man was named Örn (Eagle). He was a vagabond and skilled in black magic. He lay in wait for Thórir in Hvinverjadal when he was going south across Kjöl and made a bet with Thórir as to whose horse would be first, because he had quite a good horse, and each of them put up a hundred in silver. They both rode south across Kjöl until they came to a plain, which now is called Pigeon-nose's Course, but the difference in speed was so great that Thórir met Örn in the middle of the course. Örn was so dissatisfied with his financial loss that he did not want to live and went up to the mountain, which is now called Arnarfell (Eagle's Mount), and there took his own life, but Fluga was so much out of breath that she was left there. But when Thórir came back from the Althing, he found a white horse with a gray mane with Fluga, and with him she had conceived a foal. She gave birth to Eidfaxi, who was taken abroad and who killed seven people in one day at Mjörs (in Norway) and then was killed himself. Fluga was lost in a bog at Flugumýri (Fly's Swamp)."

Horse-fights

Numerous episodes in Saga literature show that the horse-fight was a popular amusement at public gatherings during the Commonwealth period in Iceland (930–1262) and also in Norway. A horse-fight is in itself an ugly game, and the stallions fight with fierce cruelty. Where horses are semi-wild, as they were in Europe in olden times and still are in Iceland, spontaneous fights between stallions for the groups of stud mares are common every spring. Our ancestors staged this impressive event and seem

to have derived a great deal of enjoyment from it. Frequently the fight ended with the death of one of the horses or resulted in horrible mutilations. It also happened that the horses maimed or killed attendants at the fights, and sometimes the spectators' enthusiasm reached such a pitch that a fight broke out between the owners, each with his own adherents. The Sagas contain some graphic descriptions of such incidents.

The unique *Sturlunga Saga,* dealing with contemporary events in the thirteenth century, describes a fight between two Icelandic horses that had been sent as gifts to Norway during the reign of Hakon the Old: "One summer it so happened, as often does happen, that horses were made to fight. A man named Gautur of Mel, of noble family and demeanour, was a great friend of the Sturlungs and had received a good horse from Sturla, and many people were of the opinion that it was the best horse in Norway. A man named Árni Óreida, an Icelander, had sent the king a horse, which he called the best horse in Iceland. These two horses were to be made to fight. A great number of people gathered. But when the horses were led forth, both of them seemed fine. They were turned loose and attacked each other fiercely; it was a most forceful fight, both long and severe. But as the fight progressed, the king's horse showed signs of fatigue, and the king was displeased. It was clear that the king was anything but pleased. Now Gautur walked about the circle formed by the spectators and looked up brightly with his only eye. Aron was nearby and a man with him called Thórarinn, his cousin. They did not like seeing the horse defeated. Aron was a friend of Árni Óreida's, but no friend of Gautur's. He thought he knew what was wrong.

"Now when they saw that the king did not heed his horse, they walked up to him. Then Aron said, 'Do not discredit the horse, Sire, because it seems to be the most exquisite of horses, but it does not have the conduct it is used to.' 'What conduct is that?' said the king. 'In Iceland a man accompanies each horse when it is made to fight,' said Aron, 'and has a stick in his hand, pats its rump and thus supports the horse when it rises on its hind legs.' 'If you think you can make the horse firm, Aron, go to it,' said the king. And now Aron and Thórarinn took off their coats, took billets in their hands and then walked to the king's horse, where it stood at the outside of the circle, and touched it with their billets, but it pressed on, as if it knew why they had come, ran to Gautur's horse, and the other one against it and attacked it fiercely. Now Gautur's horse had great difficulty because the king's horse was supported by force and it was commonly considered that such horses had an advantage. But as the day progressed, Gautur's horse grew tired but yet it would neither retreat nor flee. Aron and his assistant drove their horse harder until Gautur's horse threw itself down short of breath and exhausted and never stood up again."

This narrative shows that horsefights were considered excellent amusement, and also that in Iceland stallions were trained for this sport, as the horse was prodded on with a stick. Other accounts indicate that the games were organized; for instance, in one place it is stated that the horses fought fiercely for eleven rounds, and special referees are also mentioned who decided which horse was the winner.

For the breeding of the horses, the fights no doubt played an important role, since a good fighter was considered to be of a very special value. In one case it is said of a four-year-old horse that he had not yet taken part in a fight, indicating that horse-fights

were at one time the rule for all really good horses.

The annual two-week session of the Althing at Thingvellir, was not only the political and legislative institution of the country, but also a kind of national festival with all kinds of games and entertainment, horse-fights, horse trading and horse shows. At the millennial celebration of the Althing in 1930, one of the more interesting spectacles at Thingvellir was a horse-fight. There were also local assemblies (*Things*) in the various districts, in spring as well as autumn, where the horse also had its legitimate part to play. Perhaps the most peculiar kind of gathering in those early times was that going by the name of "horse-fight", a public meeting with all manner of entertainment, where the horse-fight seems to have been the main attraction, used as a bait in much the same way as dancing or bingo is today. Such gatherings were finally forbidden by a synod of the clergy in 1592, but seem to have persisted at least until 1623 when the last recorded "horse-fight" took place.

The European Horse

When Iceland was settled in the ninth century, the newcomers brought with them horses from western Norway and the British Isles. Those strong and enduring animals, with their small, muscular and sturdy bodies, were of great value to their masters in peace as well as warfare. For at least 800 years, no horses have been imported to Iceland, so that present-day Icelandic horses are not only the same race as those of the Viking Age, but probably also descended from the horses used by the legendary brothers Hengist and Horsa when they invaded England in the fifth century. (Bede says that these two Anglo-Saxon kings were the great-grandsons of Ódin, and it is highly significant that their names mean literally "stallion" (*Hengst* in German) and "horse." All this would obviously also make the Iceland Horse a descendant of Grani and Goti of heroic fame, whose historical owners lived in the fifth (Gundicarius king of the Burgundians) and sixth (Siegbert king of Metz) centuries, though the Sleipnir pedigree may be doubted!

The one-thousand-year isolation of the Iceland Horse has preserved in him some of the peculiarities lost in other European races over the past five centuries or so. Among these are the five different gaits to be discussed later. Many people think that these gaits are some fairly new Icelandic equestrian tricks, but this is far from being the case, as the names for the most peculiar gait, *tölt*, and for the horse using such a gait, *töltari*, are of Germanic origin. *Zelter* was an agile and smooth-moving horse in medieval Germany. The word also has its counterpart in Latin, for gently moving mares in imperial Rome were called *thieldos*, and there was considerable contact between the Germanic peoples and the Romans during the second and third centuries of our era.

But the *tölt* has an even older history. Greek artists of the fifth century B.C. decorated the friezes of the Parthenon on the Acropolis with cavalcades of horses riding to do honour to Pallas Athene. There we observe men riding on *tölting* horses of similar size as the present-day Icelandic horse. The method of sitting is typical, and there are clear examples of the *tölt* in the movement and the position of the horses' feet.

More than a thousand years later, a statue of Charlemagne (742–814) was made, where the great Roman Emperor and friend of the Pope is sitting on the equivalent of a present-day Iceland Horse, that is to say the original European horse. The statue is to be seen at Aix-la-Chapelle.

One may wonder how this breed of horses became extinct in Europe, but was preserved in Iceland. One major reason was the development of roads and military requirements over the past four centuries. Until the seventeenth century, the European horse was an all-round small family horse: a riding horse, a pack-horse, and a draught animal, owned and used by rich and poor alike. It was a natural part of the household, kept near home. When extensive roads brought carts and carriages, only one kind of horse was needed: the draught horse. When at the same time the cavalry turned from loose to closed formations, a special kind of horse was required. The breeds were strengthened by pedigree and crossbreeding in order to make the horses into heavier and stronger draught animals. The European peasant became poor and could no longer own riding horses. This was especially the case during and after the Thirty Years' War (1618–48). After that, it was only the army and the nobility that used horses for riding. Peasant horsemanship disappeared in all parts of Europe except in Iceland, while riding was influenced by the needs for cavalry horses, so that all kinds of racing, steeple-chasing and other tricks were introduced in order to train the horses.

However, there were two groups of

A new-born foal tries unsuccessfully to get on its feet and take its first steps.

starvation that prevailed in Iceland through many centuries. It has been shown by measuring bones, found in Gokstad and Oseberg in Norway and also in Denmark and Germany, that the Iceland Horse is the same kind as that used by the Vikings a thousand years ago, and that he has a much longer history. Skeleton finds in pre-Christian graves, when compared with similar finds in Viking graves in Norway and Iceland as well as with skeletons of present-day horses, prove the relationship between the European horse and the Iceland Horse.

There is a well known account in one of the medieval Sagas, the *Thorskfirdinga Saga,* of a chieftain, Gold-Thórir, having received a horse as a gift from Gautland on the west coast of Sweden, which in the Saga is called "the runner from Gautland". The horse was piebald and called Kinnskær. He is described as being bigger than other horses and being fed on grain both in summer and winter. The son of this horse is also mentioned in the Saga and described in similar fashion. Eastern horses were known at that time to exist in Scandinavia, and the description makes clear that this in an Eastern horse. Other imported horses may have been of Eastern origin or cross-

people in Europe in the seventeenth century that were not interested in the new breeds of horses: the emigrants to North America and South Africa. They brought with them the old breed to North America and found a similar race in South Africa (the Java Horse). Thus North American and South African horses are five-gaited.

The Iceland Horse

Obviously the isolation of Iceland and the extremely adverse conditions played a major role in keeping the race pure and developing the agility, toughness and resistance necessary for surviving the conditions and the treatment which the horse had to suffer as a result of the semi-

16

(*stód*) consecrated to a certain god. Thus the horses were bred for the pleasure and glory of the gods.

It is clear from the Sagas that the horses were selected according to quality and colour, and that each farmer liked to have his own particular colour in his horses, a tradition still alive in Iceland. It is, however, not long since foreign horse buyers, who were frequent visitors to Iceland from the last century onwards, complained to the Icelanders of the variety of colour in Icelandic horses and advocated a selection of a certain colour for the breed, as was customary in breeding in other countries and made the horses more marketable. Today most people are glad that Icelandic farmers rejected this dogmatic approach which is nowadays considered useless. Now horse-owners are concerned about ensuring that every horse possess some peculiarities, have a distinct personality, not be fashioned in a cakemould in some studbook office. It is noteworthy that Icelandic farmers have consistently refused to breed horses in accordance with market demands, but have tried to preserve with utmost precision and care the original traits of the horse, to suit the age-old traditions. By so doing they lost some of their markets both in Britain and Denmark. The

breeds, although no other accounts on that subject are to be found. This mixture proved to be fateful for the Icelandic horse breed, causing illness and degeneration for centuries. It is indeed instructive for contemporary specialists in breeding that examination has revealed that, after more than 800 years, evidence of such crossbreeding is to be found, in spite of the most severe and merciless natural selection imaginable.

Breeding

In the Sagas we find some remarkable accounts of the improvement of horse breeds, while there are no references to selective breeding in other livestock. Four to six mares were selected, together with one stallion, and the group was called *stód*, which is the same word as "stud" in English and closely related to the German words *Stute* (mare) and *Stuterei* (a stud, or a group of stud horses and mares). The concept is ancient and unchanged since the improvement of breeds was a heathen religious practice. The best horses were selected and the whole group

17

Stud horses grazing in a typical Icelandic summer landscape. The clearest signs of spring and early summer are to be seen in the demeanour, movements and vigour of the studs in the pastures.

Norwegians served the market demands better by breeding a good horse for the small farmer, uniform in colour and build, and the buyers neither asked for a spirited nor a gaited riding horse.

So far as is known, the purity of the Icelandic horsebreed has no equivalent in other breeds. The Althing (oldest parliament in the world), already in its first year, 930, passed a law forbidding the import of horses — but as we have seen, there seem to have been occasional violations of that law until about 1100. Conversely, any horse leaving Iceland can never return.

There are many references to breeding, studs and stallions in the Sagas, but from the thirteenth to the eighteenth century our sources are rather reticent on the subject. One of the few notable exceptions was Bishop Stefán Jónsson of Skálholt (1491–1518), who had studs and at his death left no fewer than 180 horses of his own in addition to the episcopate's horses. The wirst written admonition to the Icelanders with regard to better breeding practices was an essay by Ólafur Stephensen, one of the highest officials of the country, published in 1788, with prescriptions for correct breeding and descriptions of the well-built horse. His son, Magnús

Stephensen, also a high official, wrote about horsebreeding in 1825, advocating the careful selection of stallions and stud mares. The first practical steps in this direction were taken in Skagafjördur in North Iceland, the most famous horsebreeding district of Iceland, where the district authorities in 1879 appointed a three-man committee in each parish to supervise improved breeding methods. The following spring, a show of livestock took place in the district, where a stallion and a stud mare were awarded a prize.

The first law about horsebreeding was passed by the Althing in 1891, and ten years later an amendment was added where it was forbidden to let stallions older than one-and-a-half year roam freely with other horses either on the commons or the home-pastures. In 1917, 1926 and 1929 the law was revised by the Althing with a view to defining more clearly the duties and obligations of those engaging in horsebreeding.

The planned breeding of the Iceland Horse has mainly been based on the quality of the gaits. Stallions and mares are selected according to strict standards and, when they fulfill the requirements, are entered into the pedigree. What mainly characterizes the Icelandic selection of breeding horses is the rule of displaying the horse under the rider, after it has been subjected to a judgment of its external appearance. Furthermore, the horse has to be ridden by one of the judges before the final decision as to its breeding value.

The Iceland Horse is uncommonly robust, healthy, enduring and weather-resisting. He is highly versatile, makes small demands regarding feeding and housing, lives to a ripe age, and has an interesting and highly individual character. He has been described as a first-class family horse, equally suitable for all age groups of both sexes. He is intelligent, patient, easy-going, but does possess a fiery temperament when the occasion calls for it. To those concerned about appearance, he may be a little too shaggy two-thirds of the year, when wearing his "winter coat", but he still looks attractive. For all his individuality, he is a social animal, preferring to be housed with other horses and to graze in a herd.

An outstanding authority on horses from all parts of the world, the German writer Ursula Bruns, has in one of her more than a dozen books on horses (two of them exclusively on the Iceland Horse) stated that after riding all possible breeds of horses for fifteen years before the Second World War, she decided at war's end again to buy a horse. By that time, she was living in a big city, did a lot of travelling and could not stand the idea of other people riding her horse. She wanted a horse that did not interfere with her life style and rhythm, was independent and self-sufficient. What horse could satisfy such demands? After much searching, she at last found an Iceland Horse, an animal that not only met all her requirements for temperament and individuality, but was also endowed with that extra quality so rare in European horses, the fifth gait of *tölt*. She kept her newly found horse in stables where her colleagues kept some thirty larger horses and made fun of her small one. That did not bother her, and after more than twenty years of living together with her Icelander, she had no regrets and had indeed in the meantime acquired a female as well. Both these horses she describes as some of the "most explosive types" she ever came across.

Studs

Even though Iceland lies on the verge of the habitable world, skirting the Polar Circle, its climate is temperate owing to the Gulf Stream, so that about 50 per cent of its 45,000 horses live out of doors all the year round in

The foal has reached the age of walking and running. The breeder observes all its qualities, its bearing and ambling pace, the position of its head and the carrying of its legs, so as to find clues to its potentialities.

a semi-wild state, grazing on uncultivated pastures, often without needing any fodder at all in normal winters. Some farmers, however, have open barrels with herring for their studs in winter. The horses only take a certain portion each day, usually one herring. Herring has the same nutritional value as oats, but more vitamins. Thus the horse is a peculiar and outstanding feature of the Icelandic scenery to an even greater extent than are the sheep. He is found all over the country — in the mountains during summer, near the highways and around the farms in winter. He can live without causing much bother to his owner, and he does add colour and life to an otherwise cold and stony winter landscape. And the clearest signs of spring are to be seen in the demeanour, movements and vigour of the studs in the pastures. A mature stud horse is a paragon of strength and beauty. He is noble, imperious and cruelly jealous of his dominion. The mares are submissive and compliant to him. If newcomers presume to enter his turf or tamper with his authority, he defends it with fierce determination, using both jaws and hoofs. He whinnies savagely, runs amok, rising on his hind legs beats his adversary's body with his front legs, runs alongside his adversary and digs his teeth into the other's sides and neck, suddenly turning around and kicking him with all the terrible force of his hind hoofs. Such encounters often led to death or mutilation of the intruder.

The stallions are usually full of play and vitality, their bodies slim and vibrating with life, their necks raised, their ears pricked up, their eyes clear and alert. These horses have lived with the elements all their lives, looked from high mountain ridges over deep valleys and wide treeless plains or sandy wastes, sighted the mountain tops and the glaciers in the distance. Their gaze reflects something of the almost unlimited vistas of the Icelandic scenery. They are familiar with lava fields, stony hills, heaths and swamps, serene summer mornings and long dark winter nights. They exhibit a kind of philosophical detachment in adversity, being used to supporting themselves in all kinds of circumstances, to bringing their young ones into the world without any help, to looking for the last grass of summer in the farthest reaches of plains and valleys, to digging for a tuft of grass through snow or ice in midwinter, to waiting for days or even weeks for a snowstorm to subside, patiently standing under a cliff or a wall, their shaggy coats covering and concealing their supple bodies. The surefootedness of young as well as old is proverbial: they will traverse any kind of terrain at unbelievable speed.

Running Away

A well known and sometimes rather inconvenient trait in some horses in Iceland has been a strong tendency to run back home, when they have been sold or transferred to new and unfamiliar places. Such horses seem to balk at no obstacles, but will run for days, weeks and even months across mountains, wastelands and dangerous rivers to get back to where they came from, sometimes traversing half the country. There are even puzzling examples of horses being transported long distances in the holds of ships and yet being able to find their way back across the country. This homing instinct and this keen sense of direction defies any explanation and is one of the mysteries of nature.

Hobbles were made of horse-hair and the legbones of lambs and fastened to the front legs of horses as a precaution against their running away. This device was not only used on horses from distant districts that might want to run back, but also when travelling across the country so as to keep the horses around the camp.

21

Hardships

Although Icelandic winters in this century have been relatively mild and not very rough on the horses, this has not always been the case. It was indeed rare that horses died from cold in Iceland, but through the centuries there have been occasional hard winters when horses starved to death — as did their masters, too. Provisions would run short in late winter and death would be imminent if the weather was harsh. Volcanic eruptions also caused havoc time and again, sometimes killing off a third of the human population along with much of the livestock. As an example one can take the late eighteenth century, when the greatest volcanic eruption in historic times anywhere in the world took place in the south of Iceland in 1783. In 1760 there had been 32,000 horses in Iceland, but in 1784 there were merely 8,600. Accidents were also frequent owing to bad weather, for instance when blinding snowstorms caused horses to walk off cliffs or into ravines or turbulent rivers. There were no bridges in the country until this century, so that all the big and dangerous rivers around Iceland had to be traversed on horseback, which always was a tricky affair and often proved fatal to both horse and man.

Until recently it was common to keep horses in uninhabited islands for winter grazing, especially if these had rock formations or some other natural shelter in bad weather. If the island was far from the mainland, the horses would be transported in boats, one or two at a time, but for shorter distances they would swim across a sound or a strait, being towed by a boat, five to ten at a time. This writer recalls it as one of his more exciting childhood experiences when he helped tow horses across the strait from Reykjavík to the island of Videy in late autumn and bring them back to the mainland in late spring.

The Most Useful Servant

It is by no means an exaggeration to maintain that for more than a thousand years the horse meant more to the Icelanders than perhaps to any other nation (the American Indians and the Mongolians may be the only exceptions). The horse was ever present, in joy as well as grief, and had many vital tasks to perform. He bore the brunt of the long and arduous working day, and he took part in festivals and other public gatherings. Even attending church was almost unthinkable without horses. The horse was, in effect, the only means of transport in a rugged and mountainous country without roads or bridges, carrying people from one corner of the country to another, transporting goods between the remote inland farms and the trading points on the coast, bringing home the hay from distant fields, rounding up sheep from the mountains in autumn, conveying the midwife to an expectant mother or a new-born baby home from the midwife, sick people to the doctor and the dead to the grave. These were only a handful of the diverse tasks entrusted to the beloved horse, who went by the honorary title "the most useful servant."

Wheeled vehicles only made their appearance in Iceland at the turn of the century, so that the horse played the roles of cart, wagon, buggy, train and so on. Most of the transportation was done with the aid of pack-saddles which were fitted out with all kinds of ingenious devices, so that even coffins could be carried on horseback. For longer journeys between different parts of the country, and even for bringing hay home from distant fields, there would be long trains of pack-horses, usually in a single file. The long journeys would be undertaken in June and July from all parts of the country to the fishing stations

During the mating season in Iceland, each stallion is usually run with 15—25 mares in enclosed fields, and individual mares are kept with the stallions for 3—6 weeks. The reproduction rate is on the average 95 per cent.

on the coast to procure stock-fish for the winter. These trips might take weeks or even months. The rule seems to have been that people from North Iceland had their pack-horses separate and drove them in front of them, sometimes dividing the trains into several contingents if they were large, while the people from South Iceland would tie a string to the jaw of each pack-horse and tie it to the tail of another horse, thus making a train of five or six horses, each train being led by one man. The long journeys between different parts of the country, across mountains and difficult rivers, were often hazardous, and there are innumerable stories of hardships.

There is, however, evidence from the first centuries after the settlement of Iceland that, for the transportation of very ill or wounded people, a device similar to the *travois* was employed, the *travois* being a kind of vehicle used by American Indians, consisting of two trailing poles bearing a net to hold a load. The device was pulled by a horse. In some cases, two poles with a net between them would be carried by two horses, one in front, the other behind. This was both used for very ill people and occasionally for corpses, especially when rivers had to be crossed. During winter, when there was snow and ice, sleds would be used on well-to-do farms for transporting hay from distant fields, where it had been stored since summer, and especially for transporting timber for buildings such as churches and large farmhouses.

Endurance and Swiftness

The endurance and swiftness of the Iceland Horse are both proverbial, and there are numerous accounts from all ages to confirm this. Three of the most renowned will suffice here, but such stories are legion. Perhaps the most famous and indisputably the most dear to the Icelanders is the folk tale about a condemned man named Skúli, who makes his escape from the

Until this century, the most common method of transporting mail from one part of the country to another was by use of pack horses that would traverse unbridged rivers and cope with the many other obstacles of Iceland. In connection with the 1100-year settlement anniversary in 1974, a train of mail pack horses was sent from Reykjavík to the great horse meet in Skagafjördur in North Iceland, to commemorate that age-old tradition.

Althing at Thingvellir and rides all the way to Húsafell in Borgarfjördur, some 75 kilometres, on his single horse, Sörli, pursued by a group of eight men with twelve horses that never catch up with him. Having saved his master, Sörli collapses and dies at Húsafell where he is buried.

One of Iceland's most beloved poems, written in the late 19th century, deals with this dramatic pursuit.

Another well remembered ride was that of the young Árni Oddsson, later one of his country's two leading officials, from the east coast all the way to Thingvellir, a distance of some 400 kilometres, to bring a helpful message from Copenhagen to his father, Bishop Oddur Einarsson (1559–1630), in his bitter dispute with the Danish Governor-General. He rode the stretch in four days on a single horse and saved the situation for his father at the very last moment.

24

The third account is from the middle of the last century and tells of a farmer, Bjarni in Melrakkadal in North Iceland, who drank heavily and was rather unruly when away from home. His wife had fed his horse, Skörungur, on dairy products from the beginning in order to strengthen him. The last autumn of his life, Bjarni was with a train of pack-horses in the south (Álftanes) and had Skörungur alone for riding. While he was packing the horses, he suddenly felt sick and faint and only wanted to get back home before taking to bed. He asked one of his companions to take charge of the pack-horses, and then set out on Skörungur alone about three o'clock in the afternoon. Around five the next morning, he reached Brunnar south of Kaldidalur in the central highlands, where he found Jón Thorarensen of Vídidalstunga and a boy of thirteen with eight horses. They were asleep and Bjarni woke them up. Jón told him that they would set out very soon, since they intended to reach home that same evening, but dissuaded Bjarni from joining them, since they had many horses and were in a great hurry. Bjarni told him that he would not detain or delay them, even though he joined them, and that he would fall behind if Skörungur got

tired. They now set out and rode very fast. Toward evening they reached the northern edge of the highlands and descended into Vídidal in North Iceland. Now Bjarni rode up to Jón and took leave of him with a handshake, for he could no longer wait for him, he said. He rode ahead and the others could not catch up with him. Bjarni reached his home during the night after riding some 250 kilometres in one stretch and died shortly afterwards, but Skörungur suffered no harm.

The veracity of these three tales and many similar ones was brought out in the Great American Horse Race and the Pony Express Race from New York State to California during the summer of 1976, both of which were part of the Bicentennial Celebrations, where Icelandic horses distinguished themselves, as we shall see later in this book.

Favourite of Poets

There can be no doubt that the horse has been one of the most common, if not **the** most common, subject for Icelandic poets through the ages. Some of the most memorable and best beloved poems of Icelandic literature deal with the horse, especially as the companion and peer of his master,

and there are innumerable four-line stanzas (the national literary form) celebrating the many different qualities and virtues of man's most useful servant and dearest friend. Many great poems also celebrate the magnificent experience of riding amid meadows and mountains, enjoying the splendid panoramas, the vast silence and the engrossing solitude of Iceland.

Most Icelanders have been firm in their belief that horses have a life in the hereafter no less than humans, and that good horses will meet or be met by their owners beyond the grave. In ancient times, the goddess of death, Hel, rode a horse, and many folk tales relate how ghosts and other revenants rode horses around the countryside. One of the most famous folk tales actually deals with a riding ghost.

But there are also poems dealing with the tragic lot of those horses that were sold abroad during the nineteenth century and up to the Second World War to work in dusty and dark coal mines in Britain. By the thousands, the freedom-loving horses of the Icelandic mountains had to suffer the horrible voyage across the ocean in overcrowded freighters and then life-long bondage in the bowels of the earth.

Winter Riding

Today the horse has only one — and dwindling — practical function in Iceland, that of carrying the farmers in the sheep round-up in autumn. Even here, aircraft have in some instances taken over.

However, while the horse has been eclipsed in the countryside by jeep, tractor and mechanized hay-making, he has enjoyed a steadily growing interest and importance as a source of pleasure. As a matter of fact, there is probably no sport more congenial to weather conditions in Iceland than horseback-riding, for it can be done in rain or snowstorms just as in sunshine and summer heat, provided one is suitably dressed and mentally prepared for any eventuality.

In Reykjavík there are hundreds of regular horseback-riders, each owning two, three or more horses; similar interest is being shown in other towns and, for that matter, also in the countryside itself where the rearing and taming of horses has become a lucrative occupation.

Riding clubs in Reykjavík and the larger towns provide their members with stables on the outskirts of the populated areas, with easy access to open country for longer or shorter riding tours. The town-dweller has a choice of two ways to tend his horses.

Either he entrusts them to the care of a club employee who attends to the daily feeding and grooming, or he does all the work himself. Usually two or more horse owners hire or build a stable together and tend their horses in rotation, one coming in the morning, the other in the evening. The daily communion of horse and owner is to be preferred, if at all possible, for it establishes the kind of intimate fellowship which is half the enjoyment of owning and riding a horse. The daily care of horses need not be very time-consuming, but it is important to groom and talk to the horse as often as possible. As far as possible, the horses are allowed to move out of doors, both in order to strengthen their health and to preserve their contact with nature.

A horse needs fresh air, motion and tumbling or rolling on the ground all the year round. Even if the stables are well ventilated, which is very important, fresh air is necessary for the wellbeing of the horse. If he is reasonably fed, he will play around, fight and sport once he is outside the stables. This motion is healthy and necessary for his breathing and blood circulation, besides being beneficial to the muscles. Horses should, if possible, be allowed to roll on the ground or, preferably, in the snow every day,

so as to groom and harden their hide. Horses are by nature very clean, as are most animals living a great deal in the open, and they find it difficult to live in dirt or messiness.

After the active summer season, the horses are as a rule provided with rest and relaxation in autumn and well into winter, before they are housed for the winter. In December or January they are shod for winter riding, after which they are again in use two or three times a week, on weekends and one or two evenings during the week. This is to keep them fit and prevent them from adding unnecessary fat. Though the summer months are undeniably more convenient and nicer for pleasure riding, there is a lot to be said for winter riding, too. In frosty weather, the lakes will be frozen, making excellent race courses, especially for pacing, which requires an even surface. A sight not to be forgotten is a good pacer dashing across an ice-sheeted lake, let alone when a group of them race across the glazed surface. There is also a special magic attached to riding over snow-clad hills or along star-spangled beaches in moonlight with the Northern Lights dancing across the sky like gracefully draped ballerinas, not to mention the thrills of braving sleet and hailstorms,

It is quite common for riders to take their young horses along, in order to provide them with motion and accustom them to discipline and human company.

At certain intervals, the riders stop for a rest and a lively chat, compare their horses, at times trade horses, take a sip from the flask going round — it is customary to bring along some "firewater" for warming up in the cold.

cutting winds and biting frost in the company of the shaggy, surefooted horse, who in a way becomes one with his rider if they are close friends.

It has been said that to serve a horse is an art, but to enslave a horse is easy. It is mostly a question of the man's character. Though obedient and willing to serve man, almost to the point of slavishness, the horse is essentially a companion of man, demanding respect and appreciation, rendering his best service when there is this mutual respect and balanced relationship.

There are many references to horses indicating this strong bond between them and their owners. One noted lady in Iceland called her stables her "psychological laboratory". A farmer called his favourite horse "dancing sanctuary". One of Iceland's greatest poets, Einar Benediktsson (1864–1940), wrote a long and magnificent poem about horses and their emotional importance to man, stating:

The rider on horseback is king
for a while,
Crownless he possesses countries
and continents.

Many Icelandic horsemen agree with another assertion in the same poem — to the effect that alone man is only half, but on horseback he is "greater than himself". In a country so long harassed by foreign domination, abject poverty, natural catastrophes and humiliating living conditions, it was only natural that the elation of a spirited and temperamental horse was to some degree conferred on the rider, that he in a sense partook of the vitality of his most intimate companion.

It may be idle to quarrel about whether horses are governed by instinct or intelligence, but they do exhibit some very human traits, both pranks and vileness when badly treated, as well as courage, faithfulness, affection, caution and a quality which comes very close to intelligence. All the finest human properties can be found in horses when they are correctly handled.

It has been observed that good horsemen have a reputation of being

successful with the fair sex. This may be simply owing to the fact that the same kind of sensitivity, affection and patience is required for winning the heart of a woman as for gaining the confidence of a horse. Each horse is an individual with a complex character and a highly sensitive and often vulnerable emotional nature. When he meets with warmth, understanding and authority which is not abused, he

can be as sweet and pliable as a young girl in love. Which does not mean that he loses any of his temperament or unpredictability.

There is an engaging story from the early part of this century about an elderly couple who had only one horse left, a mare that was an excellent riding horse, but self-willed and impertinent. One day they went to church and had to cross a river. This

A weekend ride in the vicinity of Reykjavík.
The riders are dressed for cold weather as
well as for showers or hailstorms.

was in spring and the river had swollen. The wife rode the mare, but the husband walked along, hoping to get a ferry across the river. When they came to the river, the wife rode across it, but the husband waited for the ferry to come from the other side. But once the mare came ashore she started shaking herself wildly, so that the wife dismounted to see what was annoying the horse. But as soon as the stirrup was free, the mare ran back to the river and waded across to where her master was waiting. He mounted and crossed the river too, and there was no more shaking on the part of the mare. All the eye-witnesses agreed that her unexpected demeanour could only stem from her concern to get her master across the river along with his wife.

That mare's mother, named Sokka, was in the habit of bringing her new-born foals home to show them to her master as soon as they had seen the light of day. She did not want any help when foaling and grew very crusty if anyone tried to assist her, and she was very shy of people as soon as the ritual of showing the new-born foal was over. On the other hand, Sokka did not bring her new foals home to the farm if her master happened to have come across her in the pasture right after she had foaled.

Then the presentation was superfluous.

King Solomon says in his Proverbs: "There be three things which are too wonderful for me, yea, four which I know not: The way of the eagle in the air; the way of the serpent upon the rock; the way of the ship in the midst of the sea; and the way of a man with a maid." One is tempted to add a fifth: the way of a man with a horse.

This is not to say that women are less skilful than men in dealing with horses. If anything, they are more sensitive and proficient. The granddaughter of the woman who called her stables her "psychological laboratory" relates a story which illustrates this graphically. Early one spring in her youth, a man brought two horses to her farm. It was too early for grazing, and her father was to keep them for his neighbour until the pastures were ready. The girl was entrusted with their care, which she loved. Both the horses were quite young, one untamed, the other merely half-tamed. They were both inordinately shy of people and would not even touch the hay if somebody stood near the manger. The girl was however, not content with merely feeding them; she wanted to touch them as well. One evening she crept into the stables, but got hafl-scared when she saw the commotion her arrival created. The horses fled from one corner of the stables to the other and would have broken down the walls if they had known how to. After a long while, she finally managed to touch with her hand the horse she had been more attracted to. He shivered and shook over his whole body but did not try any pranks. She was encouraged and came closer to him with utmost care. She did not know how long it took, perhaps hours, but at last his fine head was enwrapped in her arms and his soft shivering nostrils touched her cheeks, while his brown vivid eyes looked into hers with trust and confidence. She was completely enraptured. When the horses were returned to their owner, she rode her friend with the brown eyes. The owner was a little taken aback when he saw her and asked whether that horse was not a bit too wild for her. "Rather the contrary," she replied out of spite, for the horse had been as good as she could wish for. In the autumn that same year, she rode in the company of that man from the sheep round-up, and he was riding her favourite. When they rested, the horses were turned loose with bridles and saddles, but the owner had difficulties catching her favourite again. While he and his companions were

32

running after the horse, she was busy preparing her departure, but suddenly a warm muzzle was thrust under her arm from behind. There was her friend showing her all his affection and trust, without being encouraged. She took the reins and handed the horse over to hos owner with a bad conscience, for she felt her friend had sought her assistance to be rid of his owner.

There are certain hazards involved in riding near densely populated areas with their ever increasing motor traffic. Spirited horses are not always very easy to control in chilly winter weather, and it sometimes happens that they collide with speedy cars with dire consequences. Therefore efforts are made, as far as this is possible, to keep riders off the highways. There are large tracts of unused land near most towns, and that is where the riders roam in large and small groups, each with two or three horses. They stop at certain intervals for a rest and a lively chat, compare their horses, at times trade horses, take a sip from the flask going around, for it is customary to bring along some "fire-water" for warming up in the cold.

While the horses are shedding their winter hair and before they are set free in their summer pastures, it is customary in some parts of Iceland to give them a bath in salty seawater. The horses, some of them carrying riders, are made to swim across a narrow inlet, which must be deep enough for a total immersion — a swim — though the distance need be no more than about 100 metres. The ability to swim is innate, but certain horses are better at it than others. The best swimmers are selected for riding. The rider uses no saddle, sits lightly, gives his mount free rein, but clasps his feet around his flanks, hands gripping the mane.

Summer Riding

Disregarding individual tastes, perhaps the most thrilling experience in Iceland is to spend a summer night out in the wild country amid mountains, glaciers, volcanoes, lava fields or meadows with not a single tree to disturb the sweeping panorama. The sun may be resting on a mountain top, hiding behind a range or lingering on the horizon. The eyes travel far, take in the magnificent landscape and the splendid sky — coloured like a painting by Turner. Few countries give the traveller such a feeling of vast space as Iceland, and travelling on horseback redoubles the enjoyment, for while riding it is possible to observe every little detail of the landscape simultaneously with the large vistas.

Swarms of horsemen are seen around Reykjavík and other towns and in the countryside during the summer months, both on weekends and in the evening. In addition to these regular short riding trips, longer tours into the uninhabited interior of Iceland are undertaken by small or large groups of riders. They are by all accounts the most exhilarating and memorable experience of any horseman, bringing him into direct contact with the virgin aspect of the country.

In the summer of 1974 the present writer, with four companions, took some twenty-five horses across Iceland, from Skagafjördur in the north to Eyrarbakki in the south, riding 10–12 hours a day through desolate valleys, over vast sand deserts and lava fields. Pastures were few and far apart, but the horses were in good condition, so they did not need much grazing. A jeep brought our provisions and camping equipment to the night quarters and frequently had to make long detours. Two of us rode ahead of the free horses, two trailed them, while one drove the jeep on a rotation basis. Midway we stopped at Hveravellir, between the two large glaciers Langjökull and Hofsjökull, to have a much-needed bath in a natural hot swimming pool hundreds of kilometres from any inhabited area. Two nights we spent in resthouses provided for travellers in the wilderness. All in all, this five-day trip was an unforgettable experience, strenuous and exhausting though it was.

It is considered appropriate to ride about eight kilometres in 50 minutes

Many riding clubs in Iceland organize joint one-day tours for their members, usually in spring (left).

Many parts of Iceland can only be visited on foot or on horseback. Few countries give the traveller such a feeling of vast space as Iceland, and travelling on horseback redoubles the enjoyment (right).

on longer trips and then rest the horses for ten minutes. Eight to ten hours a day are suitable stages on long trips if the grazing is good at resting places and each rider has three horses.

The major obstacles in Iceland today, as in the past, are the many big rivers that have to be forded in the interior. But obviously this was a much more serious problem in past centuries. The numerous foreign travel books on Iceland over the past two centuries almost invariably tell of the adventure of getting horses across big rivers. This is the way Charles Forbes, writing in 1860, describes one crossing: "Here we forded, but not without much difficulty; my friend the farmer and his pony having a considerable swim in one of the channels, where, swerving a few feet out of the right direction from the strength of the stream, he went head over heels out of his depth, and was instantly swept away by a four or five mile current. For some time I was uneasy about him; but sticking to his pony he eventually gained the bank a good quarter of a mile below us ..." The great adventurer and globetrotter Sir Richard Burton, in his two-volume *Ultima Thule* (1875) gives this description: "The conduct of ponies at the ferry is always amusing. They are driven in by the shouts of lads and

The scenery in Iceland is proverbially colourful. The light is strong and clear and plays on the green hues of grass and moss as well as the gray and black volcanic sand.

The contrasts in the scenery of the barren and uninhabited interior of Iceland are strong and impressive. In the moon-like landscape of lava and sandy wastes are snowdrifts, lakes and rivers (following pages).

best part of an hour, for the farmer had to find different places for us to ford it, the beds of quicksand having shifted during the day. He would go on ahead with his pony to test the ground while I, meanwhile, waited with my heart in my mouth hoping he would not disappear! He then would return, take my pony's bridle, and together we would slowly plunge through the swirling water, his daughter following with a spare pony. How our little steeds kept their feet in spite of rolling stones and the swift current, I cannot imagine, but they never slipped once and took us all safely to the opposite shore." Some 80 pages later, Miss Chapman had a different experience in North Iceland: "First we had to ford the broad Laxá river, the biggest by far that I had attempted to cross in this way. It was unpleasantly swift, and in places the water came up to my saddle. Just as we reached the middle of the river, my pony stumbled on to his knees. I clung on desperately while he mercifully recovered himself without falling right down."

Obviously, in all the travel books on Iceland written by Britons, Americans, Germans and Scandinavians, the horse may be said to play one of the major roles, being the constant companion as well as the only means

lasses, by tossings and wavings of the arms, by sticks and stones, and by the barking and biting of curs. They sidle, jostle, step in daintily, smell the water, and, after trembling on the brink for a time, some plucky little nag takes the lead. He is followed by the ruck, but there are often cowards ready to hark back: these must be forced on with renewal of stick and stone, and by driving those that have crossed up and down the bank. In dangerous narrow beds, it is often necessary to tow over shirkers one by one with a rope. The swimmers gallantly breast the flood, which breaks upon their crests; and they paddle with heads always up stream, dilated eyes and nostrils snorting like young hippopotami; the best always carry the back high. As they reach the far end, they wade slowly to shore, and fall at once to grazing." An adventurous English-woman, Olive Murray Chapman, writing in 1930 (*Across Iceland*), describes a river crossing this way: "Before reaching here, however, we had had to recross the river. It took us the

n many places of Iceland's mountainous terior, there are pens made for the eek-long autumn sheep round-up, where he sheep are kept while the area is being earched. These enclosures are often used y travellers in summer to keep their orses overnight or during rest stops and neals.

Some natural grottoes have been used for centuries both for sheep and horses. This one in South Iceland, Landmannahellir, accommodates 20—30 horses.

of transport. Here are a few random comments. Madame Ida Pfeiffer, an Austrian writing in 1852: "It is interesting to notice how the horses know by instinct the dangerous spots in the stony wastes, and in the moors and swamps. On approaching these places they bend their heads towards the earth, and look sharply around on all sides. If they cannot discover a firm resting-place for the feet, they stop at once, and cannot be urged forward without many blows." E.J. Oswald, an Englishman writing in 1882: "It is difficult to exaggerate the merits of those little animals when in their exhilarating native air, and fed on the peculiar strong grasses of the country. Such will go over ground, that would seem elsewhere almost too bad for any horse at a foot-pace, with a gay snort and a hand-gallop. As a rule, they wish to go fast wherever the roads will permit it; and unless indulged in a rattling gallop now and then, can be decidedly skittish, especially if they suspect another horse of a design of going before them." The Anglo-American poet W.H. Auden, writing in 1936 (*Letters from Iceland*): "I asked for a horse and did I get one! The farmer gave me his own, which is the prize race horse of East Iceland. He came with me and we had a marvellous ride. I didn't start too well, as when I mounted in a confined courtyard with a lot of other horses near, I clucked reassuringly at him, which sent him prancing round, scattering people and horses in all directions. I was rather frightened, but got on all right after that. The moment we got on the road, we set off at full gallop, and on the last stretch home I gave him his head and it was more exciting than a really fast car. The farmer said, 'You've ridden a lot in England, I expect.' I thought of my first experience at Laugavatn a month ago, and how I shocked an English girl by yelling for help, I thought of the day at Thingvellir when I fell right over the horse's neck when getting on in full view of a party of picnickers. This was my triumph. I was a real he-man after all."

Arrival at night quarters in the wild interior, towards evening. Nature is at its friendliest and promises a good night's rest.

The Five Gaits

The five different gaits of the Iceland Horse are unique in contemporary European horses, but are to be found in many breeds in Asia, Africa and America. These gaits, all of which are in many cases to be found in one and the same horse, are employed according to the terrain and the requirements of the rider.

The walk or step (*fetgangur*) was used when the horses were tied together in a train and had loads on their backs. It is still used by pack-horses when travelling cross-country.

The trot (*brokk*) is used when the rider is crossing rough country. It is sometimes rather hard on the rider, but actually there are many nuances of trot.

The gallop (*stökk*) is reserved for occasions when speed is the main factor, both over stony ground and grassy plains. A variant of the gallop is the canter (*valhopp*), a convenient gait when traversing mixed terrain, but it is considered unsightly and therefore not very popular amongst horsemen.

The pace (*skeid*) is used for short stretches at high speed. In pace the horse moves the two feet on each side simultaneously. In racing, the horse starts by galloping and must after 50 metres at full speed change over into pace. In the training of a five-gaiter, the last stage consists in exercising the racing pace, and the price of a five-gaiter depends mostly on the quality of his pace.

The rack or single foot or running walk (*tölt*) is the distinctive gait of the Iceland Horse, setting him apart from other European breeds. It is for taking it easy over smooth ground, and it is in this gait that the rider hardly moves at all in his saddle. The *tölt* is a gait in a quartered beat with equal intervals and the footfalls are: back left, front left, back right, front right. It is a gait which with unaltered footfall can escalate its swiftness from a mere step to great speed. One hears the *tölt* distinctly as a constant four-beat staccato; one sees the *tölt:* the horse is

proudly erect and carries the tail in a typically undulating movement. Finally, the rider feels the *tölt:* he sits, conditioned by the even four-beat rhythm, perfectly still in his saddle, without the tossing movement of the trot.

At horse shows, the participants are appraised and ranked according to certain standards. The judges give them marks from 5 to 10 for each of the attributes to be judged. There are two categories of horses: four-gaiters (where *tölt* is the main feature) and five-gaiters (where *tölt* and pace are the distinguishing marks). The scale of appraisal is divided into two categories: 40% for the exterior of the horse (subdivided into six categories) and 60% for riding properties (subdivided into eight categories).

The Iceland Horse is the only breed where exactly the same scale of appraisal is used in the land of origin and the countries to which the mounts have been exported. This scale of appraisal is also uncommonly exact, taking careful account of the purity of each gait and the character and temperament of the horse.

In order to induce a horse to show his best qualities and abilities in any given circumstance, the rider has to be in complete control. As in any other sport this requires first and

A contestant at a race bridles his horse and readies him for the test.

At horse shows and races around the country hundreds of horses are gathered and kept in fenced areas for grazing and grooming when not participating in the various events. This picture is from a horse meet at Hella in South Iceland.

Horse Shows and Races

There are 41 riding clubs in Iceland with a total membership of some 4,000 riders. These clubs form the National Association of Riding Clubs, which holds an annual assembly with representatives from all the clubs and organizes the quadrennial country-wide horse meets with shows and races, inaugurated in 1950. The seventh of these and by far the biggest was held in July 1974 in Skagafjördur in North Iceland, where no fewer than 3,000 horses were gathered from all parts of the country; out of this some 700 horses formed a colourful mass parade and some 400 took part in the shows and races. A truly magnificent event, constituting one of four major features of the celebrations connected with the 1100 years anniversary of the settlement of Iceland.

The quadrennial horse meets are organized in co-operation with the Agricultural Society of Iceland, and one of their major features is the selection and awarding of prizes to the country's best stallions and brood. The entries have been screened in advance by an expert travelling round the country, and they are shown both with and without their progeny. The

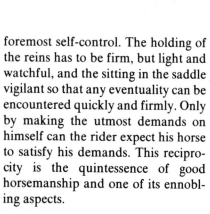

foremost self-control. The holding of the reins has to be firm, but light and watchful, and the sitting in the saddle vigilant so that any eventuality can be encountered quickly and firmly. Only by making the utmost demands on himself can the rider expect his horse to satisfy his demands. This reciprocity is the quintessence of good horsemanship and one of its ennobling aspects.

50

*Youngsters are great enthusiasts at any
horse show or race in Iceland. Here a group
of them is waiting to come forward at a
horse show.*

end, giving those interested a choice of two or three events. Betting is not practised at horse races except the ones organized by the Reykjavík club.

The criteria for judging breeding horses are international and apply to all ten member countries in the Federation of European Friends of the Iceland Horse (FEIF). The same is true of rules for riding sports; contests in the following are common in Iceland and the other European countries:

1) Tölt
2) Four-gaiters
3) Five-gaiters
4) Pace
5) Cross-country
6) Dressage

In Europe races are held in pace, while in Iceland gallop is the main feature of the races. There are five distances for mature horses in gallop: 1,500, 800, 400, 350 and 300 metres, and a 300-metre race for colts. In pace there is only one distance, 250 metres, where the first 50 metres are gallop. Recently a 1,500–2,000-metre race in trot has been introduced, and so has drilling and steeplechase. For 28 years until 1976, the Icelandic record in pace was 22.6 seconds, but then went down to 22.5 seconds, and in 1977 to 22.2 seconds.

seven best horses from each category are picked for the final competition, and these fourteen horses are then ranked by an official committee of judges using open votes.

In the years when there are no country-wide meets, the riding clubs in one quarter of the country organize the biggest horse meet of the year on a rotation basis where stallions and studmares are shown and races are

held. In addition, every one of the 41 clubs organizes its own races (sometimes two or three collaborate). A special commitee appointed by the National Association of Riding Clubs apportions the dates for the races so that they do not coincide. They are held every weekend from April through June. However, there are frequently races in different quarters of the country over the same week-

Dressed-up riders parade their prized mounts at a horse show in Iceland.

Members of a riding club lining up for the show.

One of the most famous stallions in Iceland (far left) with his progeny — the whole group being awarded prizes at a horse show.

There are three judges supervising the races and determining the order of the contestants. Then there are a number of timekeepers, three for the first horse, two for the second, and one for each of the rest. A number of supervisors are stationed alongside the race track in order to make sure that all rules and regulations are respected. Lately, a special prize has been instituted for jockeys in order to improve their demeanour and horsemanship, with good results.

Icelandic horsemanship is quite different from that of the continent of Europe. In Iceland, emphasis has been placed on developing the agility and surefootedness of the horse in

56

rough terrain, whereas on the continent the military aspect has been dominant. Icelandic horsemanship is not as uniform as that of foreign riding schools, but there is not a great deal of variation amongst the best horsemen of the country.

Since the late 1960s "uniforms" have gradually gained ground, both among jockeys and general riders, and are quite common now. Members of the Trainers' Society have their own uniform in the colours of the Icelandic flag, blue, red and white.

Since the early 1970s, some eight training stations have been established around Iceland, where experienced trainers break in young colts for owners in neighbouring districts. The Trainers' Society, founded in 1970, has about 30 members and is a conspicuous part of any major horse show in Iceland.

Usually horse shows and races are very well attended in Iceland. There were over 10,000 spectators in Skagafjördur in 1974, which probably was an all-time record. Before the Second World War, the races in Reykjavík had about them an air of a big festival with all the available buses of the capital being mobilized to transport the enthusiastic crowds the 5-kilometre stretch to the race

Thousands of spectators sit in the grass of a natural amphitheatre in North Iceland, enjoying the spectacle of their favourite horses (left). A first-prize mare with her proud breeder and cup (below).

course. There was a dance band on hand and when the races were over, the people would drink and dance into the wee hours of the morning.

The first races in modern times in Iceland were held in Akureyri in North Iceland in 1874 in connection with the celebrations on the occasion of the millennary of the settlement of Iceland. Since then, the horse has been a conspicuous ingredient of all the national festivals of Iceland. During the remainder of the 19th century, races were held intermittently in North Iceland (1875, 1880 and 1890), but it was not until 1897 that the first races were held in Reykjavík. In the following year, races were organized in various places around the country, and from then on

*Clubs lining up and preparing for a parade
at a horse show, while others are parading.
Many of the places where horse shows and
races are held have almost unlimited space
for warming up the horses and rehearsing
them before they enter the arena.*

they were an annual occurrence in **Reykjavík** as well as many other places.

The first Icelandic riding club was founded in Reykjavík in 1922, and the same summer its members finished the race course on the outskirts of the capital that was in use until the 1960s. During the following quarter-century, 12 more clubs came into being and since 1950, 28 more have been added.

The Iceland Horse matures late and should not be used or broken in until he has reached the age of five. On the other hand, he has a very long life-span and frequently preserves his faculties until the age of thirty, if he has not been broken in too early and has been well treated and fed.

The horse is connected with many highly dramatic events in the Sagas and in later Icelandic history. In Njáls's Saga, the peripety, or reversal of the action, of this great and tragic tale takes place when the hero, Gunnar of Hlídarendi, has been condemned to exile and is riding away from his home, when his horse stumbles and throws him. Looking back to his farm and new-mown hayfields, he decides to remain at home and await his destiny. This episode has been a favourite theme for poets and painters alike.

At the highest speed in "tölt", a horse carries his weight on merely one leg at a time (right).
Sigurdur Ólafsson, "king" of the pace riders, who held the Icelandic record in pace for 28 years until 1976, and Höskuldur Eyjólfsson, still training horses in his eighties (below).

Another celebrated scene, from the turbulent Sturlung Age of the thirteenth century, shows the noble and heroic Thórd Andrésson — who was betrayed by the first Earl of Iceland, Gissur Thorvaldsson, a henchman of the Norwegian king — riding to his final encounter with his adversary. Knowing he is going to be killed, he spurs his horse on as he sings in a loud voice, "My griefs are heavy as lead."

There are also horrifying stories from those early times about the dealings of men with horses. The most memorable perhaps concerns Grettir the Strong, the most famous outlaw of Iceland who was in many ways well endowed, quick-witted, strong and courageous, but had a streak of cruelty in him, best illustrated by his treatment of his father's mare, Keingála, when he was still a youngster. That episode is described in Chapter XIV of *The Saga of Grettir the Strong.*

Another noted outlaw, Eyvind of the Hills, who lived in the eighteenth century and is the only one known to have survived the prescribed twenty-year term of outlawry, was actually saved along with his harassed wife one Easter morning when a snowstorm had been raging for days and they were both on the verge of starvation. A big fat horse sought shelter under the wall of their hut way out in the wilderness. Eyvind caught him and slaughtered him, taking him for a gift from God. He belonged to a wealthy farmer who was very fond of him and set out searching for him during the ensuing summer. In September five men discovered Eyvind's hut and recognized the horse's hide hanging inside the door. The couple were caught, but managed to escape very soon.

Horses being driven into a pen before being sorted out and taken to their respective farms (below).

Other groups of horses have been driven to the pen: A lively, colourful scene. A big event in any rural district. (right).

Horses being driven into a pen before being sorted out and taken to their respective farms (below).

Sorting and Trading Horses

Every farmer in Iceland has a registered ear-mark for his horses as well as for his sheep, so that there is never any doubt about the ownership, all foals and lambs having been ear-marked in the spring. There are regular horse round-ups in spring and autumn; the horses are gathered from the mountains to a common fold in each district where they are sorted, the foals ear-marked and the ownership of all horses in the district verified.

Just like the sheep round-up, the horse round-up is an occasion for great activity; there is merry-making, horse-trading and a general survey and appraisal of the horse stocks in each district.

Horse-trading is an old tradition in Iceland and is usually practised by those who know horses well and care for them. There were, of course, also professional traders who made a living of buying and selling horses, but it was the imaginative sportsmanship of horse-trading that aroused respect and envy. Profit was not necessarily the main motivation, but rather the contest between two equals and the excitement involved in all gambling: the equal possibility of losing and winning. The trick of the trade between honourable men was not lying and cheating, but employing half-truths, innuendos, reticence and ambivalence. Somewhat like a nimble soothsayer, the good horse-trader insinuates and beats about the bush, does not deny defects in his horse, if asked about them, but never volunteers the information and prefers to talk about more agreeable matters. The bargaining itself is the essence of the game and more important than the end-result. The various choices can become quite complicated and they challenge the alertness and imagination of the traders. A straight barter of horses is no fun, for it leaves no room for a contest of wits.

*In the crowded pen, the girl has jumped
upon her horse and now tries to ride him
in cramped quarters.*

There are many intriguing stories of Icelandic horse-traders. One is about the farmer who grew tired of his wearisome hay-making, saddled his horse and rode from his farm. Toward evening, he returned, still riding the horse on which he had left in the morning, but in addition he had a number of coins in his pocket. He had spent the day horse-trading, thus making some money and enjoying it much more than the hay-making. There is another story of a good horseman in North Iceland, who was poor all his life and never owned more than a few horses at a time. He was a great expert at horse-trading and by the age of thirty he had owned about four hundred different horses.

A good horse-trader has to have the ability to size up a horse at first glance, discern his assets as well as defects, see his shape and temperament, but above all he has to be able to determine his age. The most reliable way to do this is to "read" the horse's teeth. In young horses there may be a margin of one year, depending on how they have been brought up, but usually the teeth reveal the age of a horse. First there is the dentition that occurs at a certain age, after which the wear of the teeth will tell the story from year to year. The wearing face of the teeth changes appearance with every layer that is worn away. The average wear is 2 millimetres per year. The margin of uncertainty increases with age, but the pride of every good horseman is to be able to demonstrate the age of a horse with convincing arguments.

The ear-marking of horses is not exactly a cosmetic measure, and many foreign visitors seem to object to it, to judge from comments in various travel books, but in a country where large herds of horses roam freely and many horses attempt to run back to their home districts, such a precaution is a necessity, ensuring the correct ownership of every horse in Iceland, semi-wild as well as tamed.

Sorting the wildly spirited horses in the fold is a test of manhood, requiring alertness, courage, swiftness and strength of arm as well as patience, prudence and gentleness. There is a test of dexterity between man and beasts, and perhaps Sir Richard Burton was right saying: "The domestic animals of all countries bear testimony to the character of their owners: reason, or the result of a developed brain, acts and is acted upon by instinct, or the imperfect brain produce, the two being different in quantity, not in quality. Man and beast learn to resemble each other much after the fashion of Darby and Joan: the ser-

vants of menageries, like those of mad-houses, become peculiarly brute-like, whilst animals educated by men have an unspoken language which it is not difficult to understand. In Iceland the horse has learned much from his master."

Studs and the Colours of Horses

Colour has never been a primary quality in Icelandic horses. There is a saying among connoisseurs that a good horse has no colour and that colour is only being discussed when other qualities or merits are missing. In the Iceland Horse are to be found all the shades of colours that exist in any race of horses. Approximately 20 per cent are chestnut or red, frequently with a star, a blaze and a white mane and tail. Some 40 per cent are black and bay, 10 per cent gray, 10 per cent dun (yellow and mousegray). Other more rare colours are palomino, albino-bay and albinoes. Sometimes dun and chestnut horses have a silvergray mane and tail. Mousegray horses with a silver-coloured mane and tail are a specially beautiful combination. Certain other colours change seasonally. Some of those black, bay and chestnut in spring and summer, when the coat is new, fade gradually and appear in winter to be almost white, except for the mane and tail, where the original colour is preserved.

Breeding has through the ages had a wide-ranging influence on the distribution of colours in various races. Some breeders talk of "primitive colours" (palomino and dun) and "bred colours" (red, black, etc.) and endeavour to restrict the breeding of various races to certain colours. At one time it was wrongly assumed that the piebald colour was the result of cross-breeding. Where there are many races in a country, it may be sensible to limit oneself to certain colours, for that makes it easier to avoid cross-breeds. On the other hand, a great variety of colours is a given race constitutes a genealogical treasure, and it is a matter of pride to the Icelanders that their horse stock exhibits all the colour possibilities of the species. There are in Iceland fifteen basic types of colours and colour combinations in horses. Many of these have subdivisions indicating with greater precision the particular hue or the character of a colour combination. Thus there are five kinds of red and some fifteen of piebald. As a rule, the names of horses are derived from their colours, and there is a great wealth of such names in Icelandic, in

The Kirkjubaer stud in a drove. The pictures are taken in autumn, when the horses have grown fat and stored up supplies for the winter. The similarity in colour between the tails and the withered grass is an interesting feature of the pictures.

addition to many vivid names indicating some other qualities.

Although colour is usually of subordinate importance in the breeding of Icelandic horses, some breeders have taken a liking to one particular colour and have bred whole studs with exactly the same colour. The best known of these is probably that of Kirkjubaer in South Iceland, shown in these pages, where all the horses are red with white manes and tails.

The Agricultural Society of Iceland keeps records of matings of approved stallions throughout the country. These are kept in enclosed fields,

75

*Youngsters greatly look forward to being
allowed to go to the sheep round-up, to see
the activity at the pen when the droves
come down from the mountains.*

Shoeing a horse in Iceland is a comparatively simple and quick operation. There are many sizes of horse-shoes available, and in Iceland cold shoeing is universal.

where they are run with mares owned by farmers in the neighbouring districts. Each stallion is usually run with 15–25 mares in one season, and individual mares are kept with the stallions for 3–6 weeks. Records are kept of the identification and colour of each mare, and the final record sheet on the stallion includes the sex, colour, and disposal of the foals born as the result of each season's matings. It is interesting that the reproduction rate in such enclosed fields with 15–25 mares is on the average about 95 per cent, while the fertility rate with other kinds of mating arrangements is only 40–45 per cent.

The size of the Iceland Horse is generally 131–135 centimetres stick-measure, and his witht 380–400 kilogrammes in average condition.

The Sheep Round-up

The sole practical function of the horse in Iceland today is that of carrying the farmers in the autumn sheep round-up. About mid-September, the farmers from each district co-operate in gathering the sheep from the common grazing land in the interior and drive them to a common fold, where they are then sorted and taken home by the individual owners. The sheep round-up may take up to a week, and the farmers have provisional huts in mountains where they stay at night while searching the highlands.

For this purpose, as indeed for any longer riding tours, the shoeing of the horses is a most important preliminary measure. Most farmers and most Icelandic horsemen are experts at this necessary and delicate task. A century ago (1874) Sir Richard Burton decidedly got an altogether different impression: "It is a spectacle likely to be remembered, the shoeing of Iceland ponies by the farrier, who is almost always unprefessional. Five men, without including half-a-dozen spectators and advisers, bodily engage in the task; one holds the cruel twitch, two hang on to the several limbs, one or two hold up the hoof, and number five plies the hammer. And the result is that in travelling you must always expect your animals to be pricked." If

77

Tired riders after a whole day of rounding up the sheep in the mountains. The dog has already gone to rest. There is also a woman along, which is rather unusual, since the sheep round-up is by no means easy work (below).

The people rounding up the sheep will continue until the small hours of the morning to get the droves from the mountains down to the pens. This can prove extremely difficult if the snow has come early in autumn (right).

this was true, the modern Icelander has come a long way, for it is exceedingly rare for horses to be pricked, and riders take good care to inspect the hoofs of their horses on long trips. On the other hand, shoeing a young untamed horse may be an enterprise requiring four or five sturdy men, who are sometimes quite exhausted after the ordeal. This is due to a faulty approach in the beginning since it has been shown in recent years that any colt or horse can be shod by a single person.

The sheep round-up is one of those strenuous adventures dreamed of by every youth in the countryside as a chance to put his mettle to the test. Roaming the wilderness for days in every kind of weather, hailstorms and rain as well as sunshine, demands stamina, perseverance and resource-

78

Crossing a river under some rugged cliffs in the southern part of the highlands.

ulness. There are all kinds of dangers to be avoided, bogs and swamps, ravines and crevices, torrents and crevasses, steep mountain slopes to be climbed, boulder-paved river-beds to be traversed. After a long day's search through the wilderness, a rest in the tent or the ramshackle hut is welcome, and the spirit is enlivened with a few drams of what the Icelanders call the "refulgence of the breast."

When the sheep have been rounded up, they are driven down from the mountains. It is an impressive sight to watch the droves of bleating white sheep moving slowly down the mountain slopes like waves of foaming water amid all the bustle of riders, dogs and horses. The tired and muddy riders are met by the laughter of women and children as well as other familiar and welcome noises of the homesteads. Getting the sheep into the fold entails a great deal of running and shouting. Then there is a respite before the sorting itself starts. Much of the farm population has turned up, and there is merry-making and a festive air about the whole event. Flasks pass from hand to hand amid joking and singing, and even on occasion dancing in the evening. A large fleet of motley cars surrounds the fold, while the spectators follow

the sorting in the fold with exhortations and jesting remarks.

In the countryside the sorting of sheep in the fold has an air of folk festival. In neighbouring districts the sortings do not coincide, so that there is an opportunity to attend a number of these events.

Until well into this century, the sheep round-up was a hazardous affair which frequently resulted in the loss of one or more lives. Dense fogs and terrible hailstorms on the mountains at times led to the death of all the searchers. With better weather conditions during the past four or five decades and with the use of modern technology, such as jeeps and helicopters, tragic incidents of this kind are very rare nowadays.

82

The adventures of the interior are one
thing; the great horse meets with their
shows and races are another, but the mos
rewarding experience is to be all alone
with one's horse. The horse actually takes
a person back to nature that is
increasingly becoming alien to the
town-dweller

The sunlight playing in the manes of running horses.

Man's Best Friend

It is not an exaggeration to maintain that without the horse the Icelanders would in all probability not have survived in their barren, mountainous and remote island. Not only was the horse man's most faithful and useful servant, carrying him and his belongings from one end of the country to another, but also his best friend, taking part in his festivities and celebrations, standing by his side in times of need, despair and natural catastrophes. The horse inspired our poets and painters and figured prominently in our dreams. As one writer has aptly put it, we possessed the horse and were possessed by him.

In a sense, the horse may have been the unconscious symbol of man's survival on the verge of the habitable world, standing for the freedom and intoxication of the brief and invigorating summer months amidst mountains, glaciers, volcanoes, lakes, rivers, waterfalls, green valleys and gray wastelands — and he could endure the bitter winters, often with very little aid from his owner. The horse became as much a part of nature in Iceland as a mountain or a lake, and was loved by the Icelanders with the same kind of fervour they loved their barren land.

Seeing horses at play in the wild during the summer is a stimulating and contagious experience, touching some chord deep in human nature which is constantly aspiring to some kind of harmony or unity with the elements.

Riding during the bright Icelandic summer night out into the countryside, following old paths, climbing hills, fording rivers or traversing pathless wastes, has about it an air of unreality, or rather higher reality: you may be quite alone, but you feel a strong and very personal attachment to the horse and through the horse to nature itself. You are somehow immersed in the elements, become one with the country and all those thousands of forebears, who for thirty ge-

Horses in their autumn pastures when everything is quiet and relaxed. After the last activities of summer, they are set free for a long "vacation." They get good time to rest during the dreary period of darkness, rains and snowstorms in late autumn and early winter, when the grass turns yellow (below).

The last trip of autumn. Both the colours and the light reflect the season's atmosphere (right).

nerations braved all the obstacles of a harsh landscape with the aid of the spirited and indomitable companion of man, always ready to serve, always enduring, strong and faithful. It has been said that none of God's creatures has a higher right to Iceland than the horse, which may be true, but the realistic view is that in collaboration with man the horse created the conditions which brought about Icelandic culture, which is no mean achievement.

An excellent "tölting" horse in natural surroundings in Iceland (left).

Icelandic horses in the Swiss Alps, high above the tree line and close to the glaciers. The character of these horses, their strength and endurance, captivated the Europeans. They discovered what great thrills are inherent in travelling on horseback through wild country.

Clubs and Shows in Europe

After the Second World War there occurred, not only in Iceland but in Europe as well, a radical change in the uses of the horse. He was no longer needed for agriculture, warfare or transporation. As a result he became increasingly an end in himself, that is to say a companion of man and a sporting animal. The European breeds that survived the war might be excellent sporting animals, but they were not particularly well suited to become man's constant companions. What was needed was a stock easy to keep, easy to ride, many-gaited and easy to fit into a family context. The Iceland Horse not only became the pioneer of a revolutionary development, but also found a ready market once he appeared on the scene. The amateur riders of today, who constitute about one half of all horse riders, make the same kind of demands on their horses as did the regular users of horses in former times in most countries. These demands are admirably met by the Iceland Horse. Many of those interested in or owning Icelandic horses have become experts on their gaits and uses, and now wish to acquire a deeper knowledge of their past history and traditions.

Icelandic horses enjoy the calm and quiet of a sylvan environment on the continent of Europe, so rare in Iceland (see, however, page 13).

Riding with two or three mounts was introduced in Europe when the Iceland Horse first arrived there some decades ago.

In the years 1958—1977 some 10,000 Icelandic horses were sold to families in Europe, where owning an Iceland Horse is less a sport than a particular way of life. By 1977 there were about 20,000 horses of Icelandic stock in nine European countries, where the owners have formed clubs encouraging and cultivating their use. These clubs organize annual shows and races in their respective countries, in some cases more than one a year, where the various properties of the Iceland Horse are shown and tested. At the behest of the Icelandic pioneer Gunnar Bjarnason, who in 1949 instigated the formation of the National Association of Riding Clubs in Iceland, the European clubs founded the Federation of European

Friends of the Iceland Horse (FEIF), which organizes biennial pan-European shows and races, starting in the Federal Republic of Germany in 1970.

The member countries of FEIF are: Iceland, Austria, Belgium, Denmark, the Federal Republic of Germany, France, the Netherlands, Norway, Sweden and Switzerland. Each member country is entitled to seven entries at the biennial shows and races, and these have been subjected to very severe selection in elimination events at home. Iceland is somewhat at a disadvantage in this connection, since the entries coming from there cannot return home, but must be sold in Europe along with their riding gear. Used saddles or

An Icelandic exhibition group, with renowned horses and jockeys, at the Agricultural Fair in Frankfurt, Germany This fine delegation made quite an impression at the Fair. The palomino on the right is the European "tölting" champion, while the second from the left traversed the American continent in 1976.

bridles are not admitted either. This is a necessary precaution in order to protect the stock in Iceland. For that very same reason, the European shows and races can never be held in Iceland. Obviously, the horses coming from Iceland to take part in the shows and races are coveted and highly priced.

Horses of Icelandic stock, born and brought up in a continental environment, which is radically different from the Icelandic one, are to some extent modified or influenced by the change. It is only natural that a horse who has spent his first five years in the barren wilderness of Iceland, with its almost unlimited space and freedom, should develop characteristics different from those of horses kept in fenced areas and more or less domesticated from birth.

One Icelandic peculiarity is that horses have their own set of names and are usually associated with or named for the farms from which they hail. Therefore it is quite common for European owners of Icelandic horses to make a pilgrimage to the farms of origin when they visit Iceland. The Icelandic names are for the most part (95%) retained in Europe. Indeed, the Danish Club has published a catalogue of 1,200 Icelandic horse names

Three of the Icelandic horses in the Pony Express Race during a brief respite in their trek across the American continent.

with explanations of pronunciation and meaning.

All Icelandic horses sold out of the country have their authorized pedigrees sent with them, and they are all registered. In the respective countries, state offices issue the certificates, except for Denmark where the Club has been fully authorized to do so. Thus all Icelandic horses abroad are officially registered.

In Iceland there have been almost 900 registered stallions since 1910, and about 4,000 studmares.

The Iceland Horse has participated in many national and international shows and fairs, such as *Equitana* and the German Agricultural Fair, the Swedish Outdoor Fair and similar shows in France, Norway and elsewhere. He has consistently distinguished himself for his fine and variable gaits, his endurance in cross-country riding and his agility when drilled. Courses in treating, training and riding the Iceland Horse are constantly being arranged in the various countries, with specialists (often Icelanders) travelling from one country to another.

In a very real sense the Iceland Horse has created a new breed of people across national boundaries, a group of people who have a common way of life and a common aim. We might perhaps call them a "sect", except for the fact that every one of them has his or her own particular god or gods.

Across the American Continent

Among the many spectacular events marking the American Revolution Bicentennial celebrations in 1976 were two endurance races for horses across the American continent, in both of which Icelandic horses participated. One was the Great American Horse Race, starting on Memorial Day, May 31, from Frankfort, New York, with 94 riders with two mounts each, finishing at the California State Fair in Sacramento on September 5, with 52 riders — after traversing 13 states: New York, Pennsylvania, Ohio, Indiana, Illinois, Missouri, Nebraska, Colorado, Wyoming, Utah, Idaho, Nevada and California. The foreign countries represented by entrants were Australia,

Icelandic entries in the Pony Express Race traversing the vast expanses of the Midwest on a hot day.

France, Germany, Canada, Switzerland, Austria and Iceland.

That 5,000-kilometre trek was the longest horse race ever known to have been organized. There were two riders with four Icelandic horses; one of them finished up thirteenth with both his horses, while the other came in number 21 with one horse, the other having died from some poisoned drinking water. This made him drop from number 7 to number 21, having ridden the last 1,000 kilometres on a single horse.

Another similar event, the Pony Express Race, was organized from St. Joseph, Missouri, to Sacramento, California, some 3,000 kilometres, starting on July 17, and finishing on September 15. In addition to ten riders on Arabian and Appaloosa horses, there were four riders with eight Icelandic horses, all of whom reached their destination with four of the other riders. The first of the Icelandic riders came in second in the race, the rest following him closely.

In both these cross-continental races, the Iceland Horse made quite an impression for his adaptability and endurance. The Icelandic supervisor in both races was Mr. Gunnar Bjarnason, one of the great authorities on the Iceland Horse, and his directions were decisive in securing the extraordinary results. The horses were not stabled during the race, but kept out in the open by an electric fence which was put up every night, thus making the horses "feel at home", as though they were resting in Iceland. During the hottest part of the day, long rests were taken and whenever the horses showed signs of fatigue, a rest of one or two days was ordered. In the beginning, the horses had been shod in the American fashion with "Arab plates" (as Sir Richard Burton had, by the way,

Midday rest under a canvas that was put up from the rear of a truck in order to protect the horses from the scorching sun.

proposed a century ago), but that proved disastrous, so that Mr. Bjarnason ordered the riders to revert to the Icelandic shoeing fashion, which proved to be the only right way, since the frog of the horse is used to hard surfaces of any kind.

The eight Icelandic horses participating in the Pony Express Race had already been part of the Great American Horse Race from Frankfort, New York, to Kansas City, so that they had been on the move for six weeks when they entered the Pony Express Race, which makes their performance there even more impressive. The latter followed the old Oregon Trail and then the Pony Express Trail through Utah and on to California.

The presence of the twelve Icelandic horses in the two races aroused a great deal of attention in America, and it was especially noteworthy that they proved hardier and more enduring than the Arabians taking part in the races. Many veterinarians travelled with each race, checking all the animals every day and ensuring that they were in perfect health. There was some opposition to the races all along the routes by proclaimed animal lovers, but representatives from the Society for the Prevention of Cruelty to Animals inspected the horses now and then and found them to be in perfect condition and enjoying the best of treatment.

It was quite appropriate that the Iceland Horse should take part in the Bicentennial celebrations, for his American counterparts played an equally decisive role in the settlement of the American continent as he did in the settlement of Iceland. Indeed, it has been suggested jokingly that had Leif Eiríksson only thought of bringing with him some Icelandic horses when he tried to settle in America around the year 1000, he might not have failed.

Icelandic riders in the Great American Horse Race being received by a girl with refreshments as they enter a city on their way to California.